UP ALL NIGHT!
NOCTURNAL ANIMALS

SCORPIONS at Night

Kathleen A. Klatte

PowerKiDS press
New York

Published in 2021 by The Rosen Publishing Group, Inc.
29 East 21st Street, New York, NY 10010

First Edition

Portions of this work were originally authored by Doreen Gonzales and published as *Scorpions in the Dark*. All new material in this edition authored by Kathleen A. Klatte.

Editor: Kathleen Klatte
Book Design: Michael Flynn

Photo Credits: Cover, p. 5 Vaclav Sebek/Shutterstock.com; (series background) MoreThanPicture/Shutterstock.com; p. 4 wacpan/Shutterstock.com; p. 6 Oliver Britton/Shutterstock.com; pp. 7, 9, 15 Ernest Cooper/Shutterstock.com; p. 8 Peter Yeeles/Shutterstock.com; p. 11 Nicole Mathesie/Shutterstock.com; p. 12 Kenny CMK/Shutterstock.com; p. 13 RPBaiao/Shutterstock.com; p. 17 reptiles4all/Shutterstock.com; p. 18 Kurit afshen/Shutterstock.com; p. 19 Breck P. Kent/Shutterstock.com; p. 21 Asian Images/Shutterstock.com; p. 22 Alex Stemmer/Shutterstock.com.

Library of Congress Cataloging-in-Publication Data

Names: Klatte, Kathleen A., author.
Title: Scorpions at night / Kathleen A. Klatte.
Description: New York : PowerKids Press, [2021] | Series: Up all night! nocturnal animals | Includes index.
Identifiers: LCCN 2019050083 | ISBN 9781725318816 (paperback) | ISBN 9781725318830 (library binding) | ISBN 9781725318823 (6 pack)
Subjects: LCSH: Scorpions–Juvenile literature. | Nocturnal animals–Juvenile literature.
Classification: LCC QL458.7 .K53 2021 | DDC 595.4/6–dc23
LC record available at https://lccn.loc.gov/2019050083

Manufactured in the United States of America

CPSIA Compliance Information: Batch #CSPK20. For Further Information contact Rosen Publishing, New York, New York at 1-800-237-9932.

CONTENTS

EASY TO RECOGNIZE

Most people know what a scorpion looks like, even if they've never seen one in real life. Scorpions have a long tail that turns up at the end. They also have two armlike body parts called pedipalps, which end in clawlike pincers.

Many scorpions are nocturnal. That means they hide and sleep during the day and hunt for food at night. All scorpions are arachnids. Arachnids are small animals with two basic body parts and eight legs. Spiders and ticks are also members of the arachnid family. Most arachnids are predators that hunt and eat other animals.

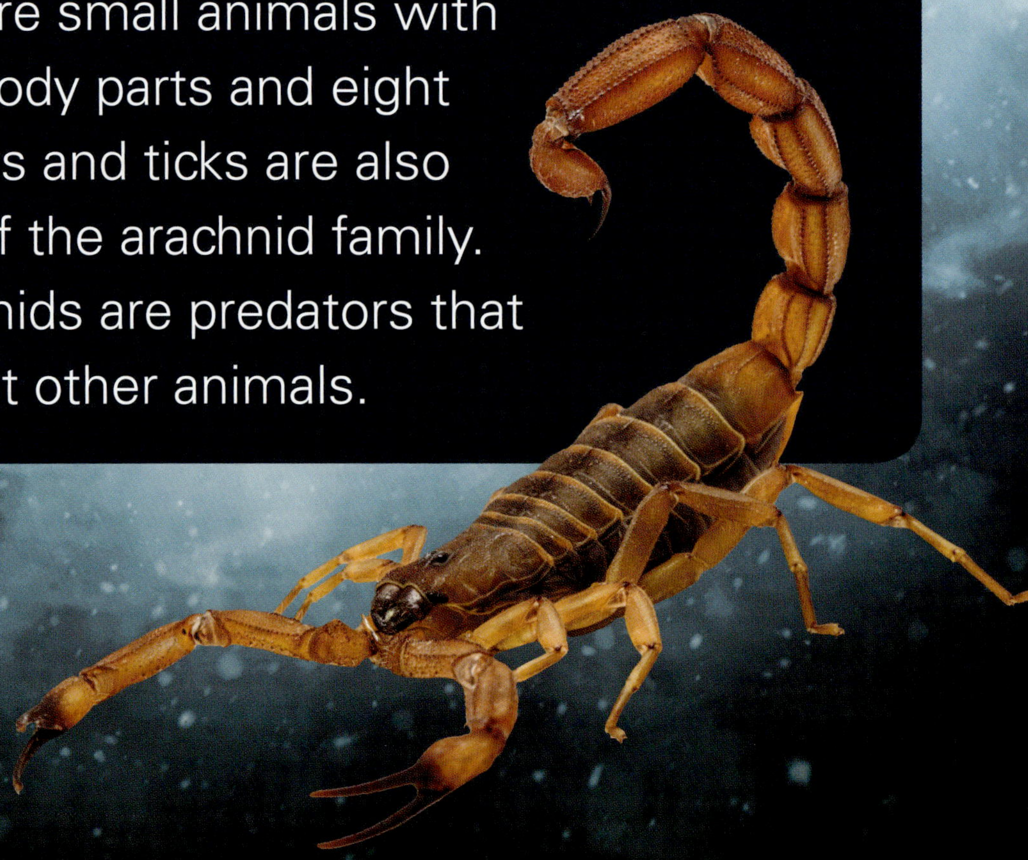

A SCORPION'S BODY IS MADE UP OF 18 SEGMENTS, OR PARTS. THIS IS MORE THAN ANY OTHER KIND OF ARACHNID HAS.

ALL AROUND THE WORLD

There are about 1,500 species, or kinds, of scorpions. The smallest scorpions are only half an inch (1.3 cm) long, but the female rock scorpion grows to be 8 inches (20.3 cm) long!

SOME SPECIES OF SCORPION HAVE SPECIAL **ADAPTATIONS** FOR LIVING IN THE DESERT. THEY CAN **TOLERATE** HIGHER TEMPERATURES THAN MOST OTHER ANIMALS, AND THEIR LEGS ARE MADE TO WALK IN SAND.

Scorpions are found in most places around the world. Many scorpions live in deserts, but some live in rain forests, in caves, or on mountains. They almost always find somewhere to hide during the day, while other animals are out looking for food. Many scorpions spend their days under stones, leaves, or tree bark. Some scorpions dig burrows, or holes, to live in, while others prefer cracks in rocks.

CLAWS AND TAILS

Scorpions have colors to help them blend in with their surroundings. They tend to be black, brown, or dark yellow. Their colors **camouflage** them, making them hard to see.

Scorpions have eyes and pedipalps on the front of their body. They use the clawlike ends of their pedipalps to catch and hold things. Scorpions have four pairs of legs. They have a part called a chelicera (kih-LIH-suh-ruh) instead of a mouth to break up their food.

MANY KINDS OF ARACHNIDS HAVE PEDIPALPS, BUT THEY'RE MUCH LARGER AND EASIER TO SEE ON SCORPIONS. THEY USE PEDIPALPS FOR EATING, MOVING OBJECTS, AND COURTSHIP.

WHILE YOU'RE SLEEPING

The first known use of the words "pedipalp" and "arachnid" was in 1826. Like many other scientific terms, they come from Latin and Greek root words.

All scorpions have sharp stingers on the tips of their tail. A scorpion's tail is curled. Along with their large pedipalps, this gives scorpions their familiar shape.

A VENOMOUS STING

One of the scorpion's main **defenses** is the stinger in its tail. A scorpion stings an animal that's trying to run away. The **venom** in the stinger makes the prey hold still while the scorpion eats it. Their venom can also kill an animal. About 25 kinds of scorpions have venom strong enough to be a danger to humans.

Scorpions also sting animals that attack them. Several nocturnal animals, such as owls, bats, and lizards, eat scorpions. Some animals that eat scorpions will break off a scorpion's tail to keep from getting stung. Some animals are even immune to scorpion venom.

WHILE YOU'RE SLEEPING

Most scorpion species aren't dangerous to people. Their stings are about as painful as a bee sting, and people recover very quickly.

A SCORPION IS LARGE ENOUGH TO MAKE A GOOD MEAL FOR MANY KINDS OF ANIMALS.

HUNTING IN THE DARK

Most scorpions depend on senses other than sight to hunt. Scorpions can sense **vibrations** in the ground. They're able to figure out how far away their prey is and how large it is. Some species can sense vibrations in the air.

WHILE YOU'RE SLEEPING

Scorpions have parts on their legs that help them sense motion. Tiny hairs can sense vibrations in the air. This helps them to move and hunt in the darkness.

BIRDS AND ANIMALS AREN'T THE ONLY ONES WHO EAT SCORPIONS. IN SOME PARTS OF THE WORLD, FOOD VENDORS SELL THEM AS TREATS.

Scorpions are quiet hunters. Many scorpions sit still in the dark and wait for prey to come near. In fact, some scorpions just stick their head out of their burrow and wait for prey to pass by! A few scorpions search for prey. When a scorpion senses prey, it will grab it and begin to eat. If needed, the scorpion will sting its prey.

A LIQUID DIET

When scorpions come out at night to hunt, they generally catch other nocturnal animals. They like to eat insects, spiders, and other scorpions. Large scorpions sometimes also eat bigger animals, such as lizards and mice.

Scorpions are excellent hunters. However, they have a strange method of eating their food. Once they've caught their prey, scorpions tear it into little pieces. A scorpion's body makes special **enzymes** that they spread on their prey. This makes the prey break down into a liquid. Scorpions then suck the liquid into their stomach. Materials such as fur are left behind.

IT TAKES A LONG TIME FOR A SCORPION TO EAT ITS PREY. BECAUSE THEY EAT SO SLOWLY, THEY TEND TO EAT A LARGE AMOUNT OF FOOD AT ONCE. THEY CAN GO WITHOUT EATING AGAIN FOR A LONG TIME.

MOTHERS AND BABIES

Baby scorpions are called scorplings. The average litter of scorplings is about 25. Some litters are smaller, while others have more than 100 babies.

Newborn scorplings climb onto their mother's back. The scorplings stay there for several weeks, using food reserves in their own bodies and getting water from their mother's body.

When a scorpling's first **exoskeleton** becomes too small, it breaks open and the baby crawls out. It then grows a new exoskeleton. This is called molting. Although scorplings generally go out on their own after their first molt, they will molt five or six more times before they become adult scorpions.

THIS IS A FEMALE SCORPION. SHE'S CARRYING HER BABIES ON HER BACK. THE BABIES WON'T SURVIVE WITHOUT CARE FROM THEIR MOTHER.

COOLING OFF

Scorpions most commonly live in hot climates. These scorpions stay cool by hiding in burrows during the heat of the day. They come out only after the sun sets and the air cools.

Hiding from the sun also helps scorpions keep water in their bodies. All animals need water. However, most scorpions don't drink much. Instead, they get water from the liquids in their prey.

SCIENTISTS HAVE FOUND SCORPION FOSSILS THAT DATE BACK TO THE SILURIAN PERIOD—200 MILLION YEARS BEFORE THE DINOSAURS LIVED. MODERN SCORPIONS HAVE MUCH IN COMMON WITH OLD ONES.

WHILE YOU'RE SLEEPING

Although they're more common in deserts now, scorpions first **evolved** as marine animals. They were very large—up to 8 feet (2.4 m) long—and are believed to be related to modern horseshoe crabs.

Scorpions don't eat often. Since they don't move around much, scorpions generally use little energy. This means they don't need to catch prey often. Scorpions store food so well that some species can survive on one meal for 12 months.

LIVING WITH SCORPIONS

Very few species of scorpions have venom that's strong enough to kill people. Scorpion stings can hurt a lot, though, and they can make a person very sick. Anyone who is stung by a scorpion should stay calm and call a doctor. Doctors have medicines that can keep even the strongest venom from hurting someone.

Still, people who live around scorpions should watch out for them. People need to be careful around rocks or woodpiles where scorpions might hide. A scorpion generally won't attack a person unless the person attacks or hurts the animal first.

WHILE YOU'RE SLEEPING

Many scorpions are solitary, but in some places, they do live close together. This means they can account for a large amount of the **biomass** in their **habitat**.

LIKE MOST WILD ANIMALS, SCORPIONS JUST WANT TO AVOID PEOPLE. GIVEN TIME AND SPACE TO ESCAPE, THIS SCORPION WILL SCUTTLE AWAY WITHOUT HARMING ANYONE.

WHERE IN THE WORLD?

Scorpions live in many places around the world. You might be able to find them where you live. Scientists observe scorpions by using a special light called a black light, which makes scorpions glow in the dark. This makes it easier to see them. Some scorpions blend into their surroundings during the day.

Many people keep scorpions as pets, but it's important to have scorpions in the wild. Animals such as owls, meerkats, and coyotes eat scorpions and need these arachnids to live. Scorpions also help control insect populations where they live. They play an important part in the **environment**.

GLOSSARY

adaptation: A change in a living thing that helps it live better in its habitat.

biomass: The amount of living matter in an area.

camouflage: Colors or shapes on animals that allow them to blend in with their surroundings.

defense: A feature of a living thing that helps keep it safe.

environment: The natural world around us.

enzyme: A chemical matter in animals and plants that helps to cause natural processes (such as breaking down food).

evolve: To grow and change over time.

exoskeleton: The hard outer covering of an animal's body.

habitat: The natural home for plants, animals, and other living things.

tolerate: To experience something without being harmed.

venom: Something an animal makes in its body that can harm other animals.

vibration: Small, quick movements.

INDEX

WEBSITES

Due to the changing nature of Internet links, PowerKids Press has developed an online list of websites related to the subject of this book. This site is updated regularly. Please use this link to access the list: www.powerkidslinks.com/uan/scorpions